A Review of Malaria in Sub-Saharan Africa

A Case Study in Liberia, West Africa

by

Rockefeller F. Cooper II

Bloomington, IN  Milton Keynes, UK

authorHOUSE

AuthorHouse™
1663 Liberty Drive, Suite 200
Bloomington, IN 47403
www.authorhouse.com
Phone: 1-800-839-8640

AuthorHouse™ UK Ltd.
500 Avebury Boulevard
Central Milton Keynes, MK9 2BE
www.authorhouse.co.uk
Phone: 08001974150

First published by AuthorHouse 5/25/2006

ISBN: 1-4259-3338-6 (sc)

Printed in the United States of America
Bloomington, Indiana

This book is printed on acid-free paper.

Table of Contents

Abstract

For more than a century, mankind has been victimized by malaria. This disease which is considered to be the number one killer of women and children affects people who live mainly in tropical areas such Southern America, Sub-Saharan Africa and Asia. Its level of effect is based upon the type of *Plasmodium* involved, which includes P.*ovale, P. vivax, P. falciparum* and P.*malariae.* The major vectors for malaria are the female Anopheles mosquito such as *gambiae, funestus* and *melas.* As a means of combating this form of terrorism whose mortality rate is even 500 times that of HIV/AIDS (on an annual bases), a lot of drugs have been developed and are still being developed and marketed as a means of replacing the most common or drug of choice but "malaria-resistant" *chloroquine.* Some of these drugs include *Malarone, Mefloquine(Larium), Lapdap, Primaquine(tafenoquine), Artemisinin-based Combination Therapy* and a host of others. The combination therapy approach appears to be quite effective. There is now an initiative to target the genetic basis of this infection as well as to develop effective methods of prevention through vaccines, education and outreach programs. This disease has its greatest toll (90%) of mortality and morbidity in Sub-Saharan Africa, which is affected by the most virulent form of the malaria parasite...*P. falciparum.*

A case study of malaria and its effect in the West African Republic of Liberia exhibits the devastation of malaria. Liberia is a country that has experienced bloodshed for more than a decade due to a senseless civil war. Concurrently, as a result of its tropical location, Liberia had to also deal with the proliferation of malaria infections since proper preventive measures were unavailable. Unfortunately, this country had to endure the harsh effects of the disease in every form since she was under sanctions as well from the UNITED NATIONS (U.N) thus preventing the importation of certain preventive chemicals and drugs due to fear that they might be converted into war materials. It is now expected that since a newly elected and democratic government took over in early January of 2006, the malaria issue will become top issue since "health" is one of the government's top priority.

Overview

Looking at the world today, there are a lot of diseases that can be found through the length and breadth of the universe. However, malaria has joined the ban wagon thus becoming a global issue on the table of health problems. As this disease affects mainly tropical areas, 10% of the global population suffers from its chilling and feverish effects. Due to this factor, more than a million deaths occurs annually as a result of this preeminent tropical parasitic disease. (14- H).

From a continental view, Africa (mainly sub-Saharan) receives 90% of these deaths in which most of the victims are a significant amount of pregnant women and children under the age of five, as received from the 500 million cases a year. In adjacent to this, malaria in Africa which is the leading cause of morbidity, mortality, and decreased economic productivity at a high burden, encourages a disproportionately high fertility rate—parents want to have more children to replace the ones they are likely to loose. This impact on the other hand can lead to investments in health and education to be minimized for each child. Due to the detrimental fatality rate of this parasitic infection, foreign investments are stagnant, tourism depressed, and the movement of labor between regions hindered as well.

The cost of executing a basic control program has increased from $2 billion to $12 billion, (an equivalent of 6% of the Gross national Product [GNP] of some African countries), since the disease causing parasite is rapidly escalating its resistance to anti-malarial drugs being used currently. (14-H). Because of its vast financial increment, which increased in 2000, African Head of States met on April 25th in Abuja, Nigeria to launch an all-out war against malaria. Also, a study in the same year published by the Harvard University Center for International Development estimated that if malaria had been eradicated 35 years ago (1965), the annual GDP for Africa would be $100 billion more than it is today—a figure which dwarfs the international aid given to the continent annually. In the said year alone, the cost of treatment and lost of production was $2.5 billion (USD). (14- H)

Finally, as a means of understanding the continent's plight, the United States government under the leadership of President George W. bush put an initiative for malaria control into motion. This program has been launched in Uganda, Tanzania and Angola and it is expected that the U.S. government will donate $9.5 million for the implementation of the program in Uganda. About fifty percent of the said amount will be allocated to the purchasing and distribution of nets and other essentials to refugees residing in the northern part of Uganda. This action by way of example initiates President Bush's pledge in 2004 to increase funding for malaria prevention and treatment by more than $1.2 billion over a five year period in Sub- Sahara Africa.

Malaria

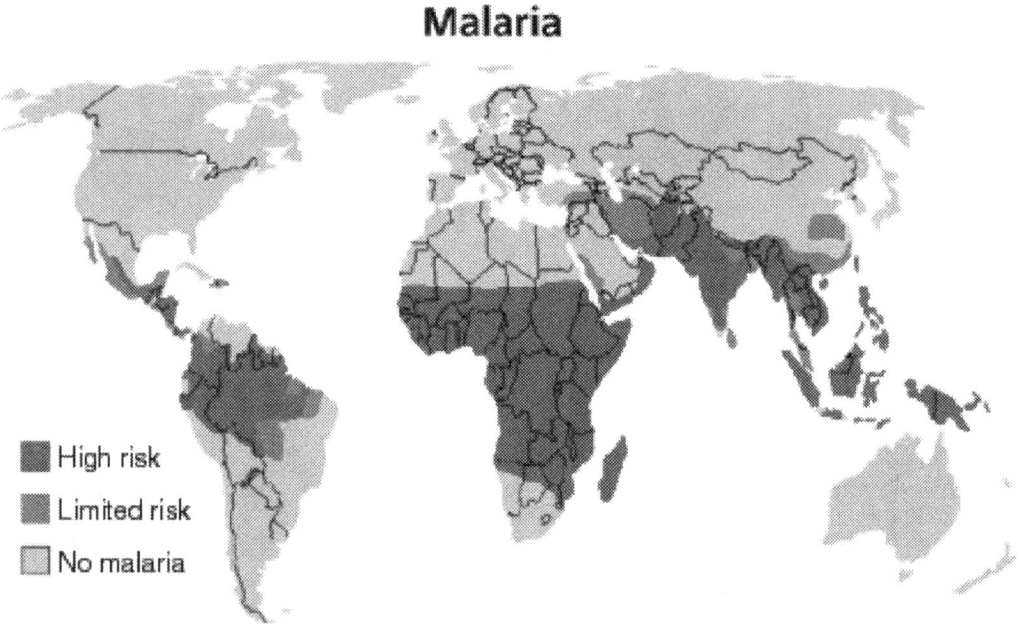

Figure 1. A global view of malaria

Malariology

This term is simply the study of malaria, which happens to be an infection by protozoa of the genus **Plasmodium**. The vector for malaria is the female anophele mosquito; and, namely, they are A. gambia (> 96% and represents A. fauna), A. arabiensis, A. funestus, and A. melas. These vectors carry parasites such as P. falicparaum (prime cause of malaria), P. vivax, P. malariae, and P. ovale, which are injected into the blood stream of the host upon being bitten. (8).

Cause of Widespread Infection:

As stated earlier, Africa accounts for 90% of death by malaria on a global scale. Though the continent is fighting rigorously to destroy this fatal act, it seems impossible despite the different avenues of support for eradication of the disease. This of course can be attributed to three key factors: ignorance, poverty, and the lack of commitment.

<u>**Ignorance:**</u> People are not fully educated or, in some cases, are completely uneducated towards the disease and its devastating outcome. As rural and suburban areas are usually attacked on a higher scale, people from these sections are highly susceptible to being victimized. This is usually the case due to poor health practices that are conducted within the communities, which have become a part of their lifestyle as these acts were engulfed and dissimilated through cultural diffusion.

By the way of example, with regards to poor health practices: **1.**—Domestic refuse are not disposed properly. Due to this, pools of stagnant water can be seen which serves as a base for the complete metamorphosis of mosquitoes—the vector of malaria parasite(s).**2.**—Dump sites are not formed at isolated areas. Since they are exposed, they serve as a breeding ground for mosquitoes. This is worse, especially during the rainy season when these garbage areas are then waterlogged. Adjacently, we can consider the misuse of public latrines. People leave urine on the floor consistently which can serve as a form of attraction to mosquitoes and a breeding ground.

All in all, this serves as another mechanism for the mosquito to feed on us humans, thus a way to spread the disease.**3.**—We can also consider areas that are swampy, bushy, or even tropical forest areas. These terrains serve as host environments for the vector since they can reside comfortably in marshy and waterlogged areas. When these situations do exist in an environment, there is bound to be a widespread of the disease due to the increment in the population of the vector.

<u>**Poverty:**</u> Due to the poor economical status of African countries, people are not able to purchase preventive equipments (mosquito spray, screens, Insecticide Treated Nets [ITN's], insecticides [DDT-limits transmission], etc.) (5). In addition to this, governments are not able to provide these necessities for their citizens due to high cost of nation wide supplies. This causes a rise in the level of infection. For instance, people farm a lot in Africa and they have to cultivate the bushes and forests into farmland. These farmers are exposed to being bitten by mosquitoes and hence malaria. Also, they live nearby the host zones of the vector, which has a high influx of the mosquito, thus giving themselves a higher exposure rate than many other African inhabitants. These factors seem to be habitual since preventive methods are not applied due to financial constraints. Furthermore, due to poverty, people are not able to purchase new drugs but instead rely on drugs that are malaria resistant (chloroquine), which is ineffective and very cheap (10 cents per pill on the average) (5).

<u>**Lack of Commitment:**</u> Messages on controlling the widespread of the disease have been disseminated through workshops, media, and even the school systems by local and foreign health care organizations, which are not implemented properly. The reasons are the following: **1.**—Those who were trained to amplify the message are not given it out in its full content due either to the fact that they were not adequately trained or they are not taking the time in teaching the general public on how to prevent or eradicate the widespread of the disease in a proper manner.**2.**—As the public is being educated, they are not practicing what is being preached to them. This may be due to the fact that poverty does not allow them to purchase preventative equipments or they might not be getting the message in its totality. This can be a result due to language barriers or the failure of the educators to fully implement an outreach program, which in turn denies the people in remote areas of the message.

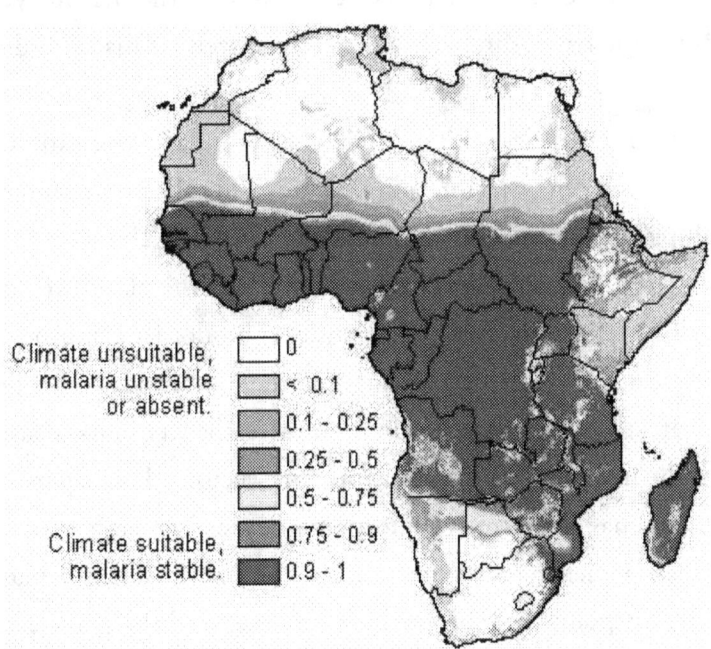

Figure 2. Map of Africa showing malaria concentration with regards to climate control.

Mode of Transmission

The main transmission vector for malaria is the female mosquito (genus Anopheles) (9). As the saliva becomes infested with Plasmodium parasites (mainly P. falciparum) after biting an infected person, the parasites can be easily transmitted into another human through a single bite. In the blood, the parasites are now known as **sporozoites** and they continue their immediate journey to the hepatic cells of the liver.

In this vital organ, these sporozoites undergo multiple fissions asexually and transform into **schizonts**. At this stage, where they contain six to twenty-four nuclei, there exist the enlargement and segmentation of these products (schizonts) into smaller mononucleated cells called **merozoites**. These newly formed cells are released where they either infect other hepatocytes or erythrocytes during the pre-erythrocytic or erythrocytic stages respectively. This is all due to the lysing of the hepatic cells.

Once inside the erythrocyte, Plasmodium undergoes a similar process to the replication stages in the liver. (9). The merozoite enlarges into a uninucleate cell called a **trophozoite** whose nucleus then undergoes multiple asexual fissions to produce a schizont. The schizont then divides and produces mononucleated merozoites, which are released into the bloodstream due to the lysing of the erythrocytes. This release also leads to the infection of other erythrocytes and eventually triggers the onset of great thirst, chills, sweating, fatigue, and fever, which produces the cyclic paroxysms that are characteristics of malaria. In most cases, the victim has continuous fever rather than periodic due to the infectious result of the primary vector—P. falciparum.

At times merozoites may differentiate into **gametocytes**, which are unable to rupture the erythrocyte. When these gametocytes are ingested by the mosquito during a bite they development into gametes. It is at this point that the erythrocytes lyses and the gametes then fuse to form diploid **zygotes** called **ookinetes**. Then there is the penetration of the mosquito's mid-gut wall by the ookinete, which then forms an **oocyst** on the external surface. The oocyst also undergoes meiosis and forms sporozoites, which migrates to and penetrates the vectors salivary gland one the oocyst ruptures. During this stage, the sporozoites are now available to be injected into the bloodstream of a perspective host thus continuing the widespread of the disease based upon numerous bites executed by the vector. Secondly the disease can also spread through blood transfusion and contaminated needles. (9)

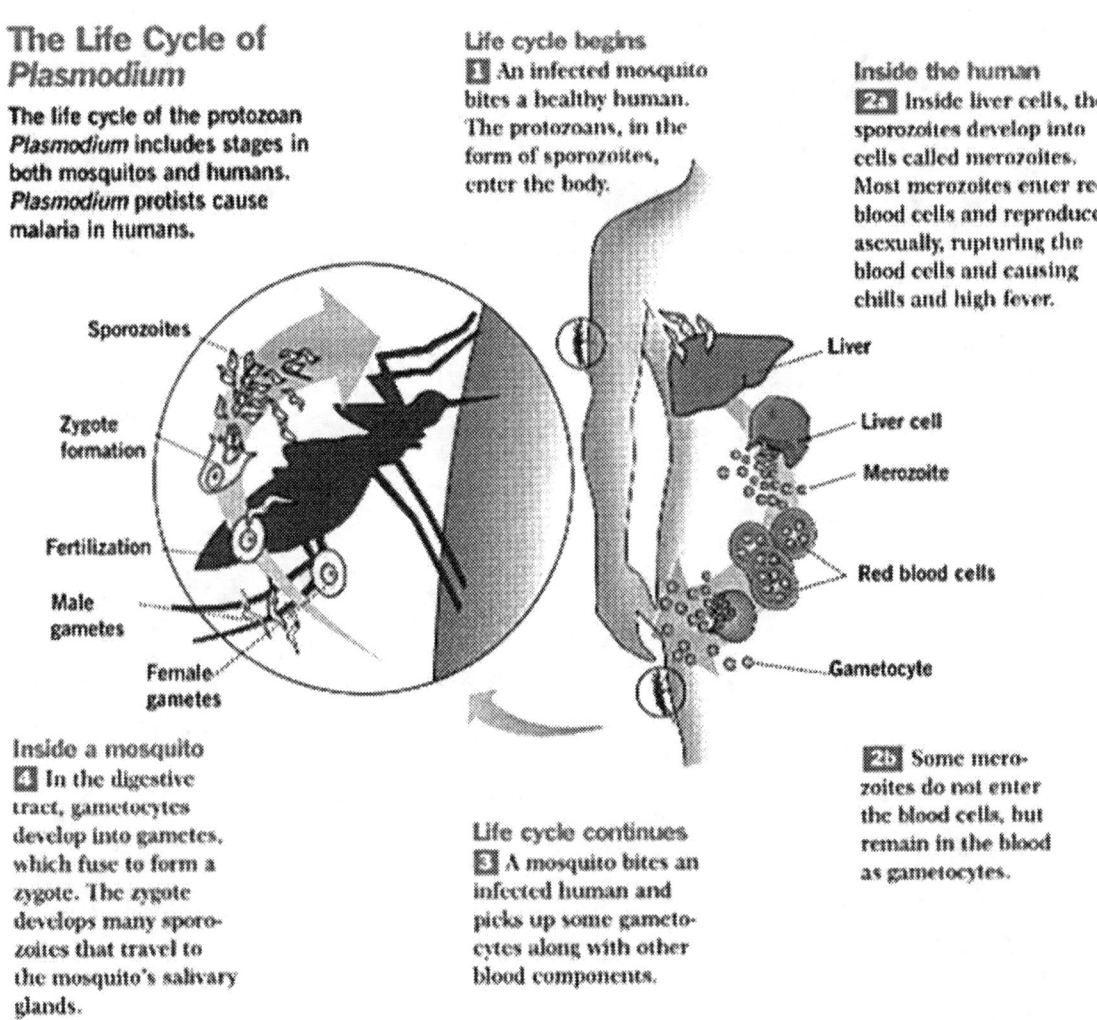

The Life Cycle of *Plasmodium*

The life cycle of the protozoan *Plasmodium* includes stages in both mosquitos and humans. *Plasmodium* protists cause malaria in humans.

Life cycle begins
1 An infected mosquito bites a healthy human. The protozoans, in the form of sporozoites, enter the body.

Inside the human
2a Inside liver cells, the sporozoites develop into cells called merozoites. Most merozoites enter red blood cells and reproduce asexually, rupturing the blood cells and causing chills and high fever.

Sporozoites

Zygote formation

Fertilization

Male gametes

Female gametes

Liver

Liver cell

Merozoite

Red blood cells

Gametocyte

Inside a mosquito
4 In the digestive tract, gametocytes develop into gametes, which fuse to form a zygote. The zygote develops many sporozoites that travel to the mosquito's salivary glands.

Life cycle continues
3 A mosquito bites an infected human and picks up some gametocytes along with other blood components.

2b Some merozoites do not enter the blood cells, but remain in the blood as gametocytes.

Figure 3. A basic diagram showing the malaria plasmodium being injected into the host through a single bite by the **female anophele mosquito**.

Personal Experience

Growing up as a boy, malaria was a major issue on my health menu. This was due to the fact that I always played outdoors which automatically exposed me to being infected by mosquito bites. This problem became very severe in my life since my infection rate on an annual basis was approximately twice every quarter especially during the civil war in my home country Liberia. I personally felt that this was catalyzed by my migration through the tropical forest from one place to the other as a means of finding safety and trying to avoid being harassed if not killed by rebel forces. In addition to this, some make shift safe havens did not have proper or no screening at all placed at the doors nor windows. This was not seen as an issue since "keeping alive" alone was then a top priority, thus creating an avenue for mosquito infiltration which led to the widespread of malaria. In adjacent to this, since the rate of infection was so high, the disease was easily transmitted from one infected person to the next through a single bite of a mosquito.

As a means of treatment, I was always administered chloroquine (tablets or injections) until I developed an allergic reaction to the drug. This left me with no choice but to seek other alternatives since my itchy reaction being so intensive, would leave my scratching and very irritated for days. On the contrary, as this allergic reaction seems to be familiar with other victims, I discovered from some of my friends and also experienced that when other alternatives are unavailable, chloroquine can be taken in a mixture. This includes the intake of chloroquine with sugar solution (one cup of unspecified concentrated sugar water for every two pills taken) or taking the drug with phenegan, being in a combinative form of oral consumption or injection. These combinations help in drastically alleviating the itchiness as well as their duration. I know this to be a fact since during my days Liberia when the civil war was at its peak; there was scarcity in certain kinds of medications. Nevertheless, other drugs such as quinine sulphate, fansidar and a host of others have seemed helpful in treating me.

Education/Enlightenment

This aspect of the malaria issue incorporates "training and monitoring." By viewing these individually, let us begin with training. In a network format, it can be viewed in four different components—training in malaria message, community-mass education campaigns, providing Malaria Disease Control (MDC) materials to all schools, observing a national malaria week for advocacy and social mobilization purposes.

By presenting this educational network, there has to be the training of health workers, care givers, pharmacy technicians, school teachers, elders of towns and villages, and people willing to receive the message on the disease and its entirety. When this is done, a mass education campaign is then said to be in place where the communities are enlightened on how to handle the widespread of the disease. The message is also disseminated through the society by exposing students within the different communities about the malaria disease and how it can be suppressed. This methodology is one of the best in a sense that when these students are home, they will be able to share the message with their families and practice them as well.

On the other hand, the media can also be very effective in this process by broadcasting malaria disease messages in different local languages as well as having theatrical plays at town halls and on television in the same format. In another perspective, free workshops can be held for the public where participants are given gifts (nets, sprays, etc...) for their participation. This will definitely attract other people and cause them to participate as well, thus increasing the spread of information on malaria control.

Finally, on an annual basis, a week should be set aside where the advocacy of malaria suppression and eradication will be of primary focus. This week will also focus on the mobilization of training groups to spread the message throughout the different countries. After training, the next stop would be to make sure that the knowledge is implemented and enforced as a means of ensuring that there is a decrease of malaria, if not totally eradicated in the African environment. This can be done by "monitoring" which can be enforced on four principles.

First, it should focus on the most afflicted regions, mainly sub-Sahara. Secondly, it should recognize that malaria control is uniquely site specific, dependent on the climate control, vector ecology and biology, and human activity. Thirdly, there should be the pursuing of two tracks: increased malaria control (both prevention and treatment) with existing technologies, together with a major investment in R & D for new technologies. Lastly and above all, the program should be funded adequately and consistently for at least 20 to 30 years if it is aimed at becoming successful.

Symptoms of Malaria

Malaria expresses symptoms within 24 hours of the first bite which are expressed in three forms.

Primarily, the victim experiences violent chills and shivering for about an hour or two. Afterwards, the individual has a vast increment in temperature maxing out to a high of 107 degrees Fahrenheit followed by high pace breathing for another three to four hours. The last form is diagnosed by profuse sweating which also last for two to four hours.

Other common symptoms include fever, joint pains, thirst, headache, enlarged and tender spleen, confusion, convulsion, coma and dark urine. As for those who are highly infected, they may experience respiratory distress. This form of distress is characterized by flaring of nostrils, intercostals or sub-costal recession and the use of accessory muscles for breathing or abnormally deep breathing. If an early or inappropriate diagnosis are not carried out, mortality, if not a poor health might bestowed upon the victim. (14-I).

Prevention and control of Malaria

As we now know, malaria is a disease, which can cause fatality in its worst case within tropical and sub-tropical areas where it has a widespread epidemic.

From both a continental and global standpoint, this disease has proven to be highly detrimental with nearly 500 million cases annually. Based upon this malaria prevention can be executing using the "**A-B-C**" method:

Awareness of Risk: Being bitten by a mosquito is very risky considering the fact that there are different types of malaria that can be transmitted depending on the location of the perspective host. This is very important especially if one is in or traveling to a region (s) that is infested with P. falciparum (> 90% of malaria cases).

B*ite Avoidance:* Most precaution should be taken both at twilight and at night. Sleep in rooms with screened windows and doors (completely screened, no holes) if you would prefer to enjoy the cool night breeze. Before going to bed, spray the room to kill mosquitoes that may have entered during the day. This should be done at least two to three hours prior to bedtime as a means of avoiding respiratory irritation due to the pungent smell of the insecticide that may also lead to rapid coughing.

On the other hand, mosquito nets instead can be used around the bed. They are highly effective, once impregnated with insecticides such as pyrethrum, permethrin or deltamethrin (2). These nets can also be used when sleeping outside once they are hole less and the ends are tucked under the mattress. During the daytime long pants, long sleeved shirts, and thick socks can be worn to avoid bites after sunset when one is moving about outside. Since it may be difficult to follow such advice, especially for those in hot climate areas (Africa), mosquito repellant containing diethyl toluamide (DEET) is highly recommended as a primary preventive method. This means that the prevention is highly affirmative once the manufacture's instructions are followed due to the chemical component of the repellent, which may cause ectodermic reactions if used improperly.

***C*hemoprophylaxis:** As it is commonly believed and said, "Prevention is better than cure." It is better to take prophylactics while in malaria infested zones to prevent being infected than catching malaria itself. This may be done on a daily or weekly basis depending on the drug in use. On the other hand, one should know the exact prophylactic to consume since different affected areas have different types of malaria. As an example to this, a combination of chloroquine and proguanil will be effective in parts of southern Asia (India) and not Africa. This is due to the fact that P. falciparum—the major cost of malaria in the continent's sub-Saharan region—is resistant to the combination, mainly chloroquine, which was effective by 70% only a decade ago and has declined drastically since then. Nevertheless, the use of Mefloquine (Lariam) is highly effective for the continent's consumption (over 90%) as well as other areas that have chloroquine-resistant malaria.

Some commonly used prophylactics include: Arloclor or Nivaquine, Proguani (Paludrine) Mefloquine (Lariam), Doxycycline and Malarone. There are also some questionnaires that one should self-encounter before ingesting any of the preventive drugs and they are: are you pregnant, breastfeeding, epileptic, taking other drugs, or is it for a child? This is to avoid health complications. As for children, their ages and body mass has to be viewed seriously so as to know the right drug to administer.

Finally, pregnant women should stay away from falciparum malaria (chloroquine resistant) areas, because the disease increases the risk of abortion, premature and still births as well as a maternal death. As for those living in these areas (sub-Saharan Africa), they can take mefloquine, which can quiet these risk factors and has no effect on the fetus. As malaria infection in sub-Sahara Africa usually spreads geographically due to existence of tropical forest and swampy areas, the use of "Bacillus sphaericus" is very necessary. This insecticide tends to subdue the larval population of the A. gambiae—the vector of the major source of malaria—P. falciparum.

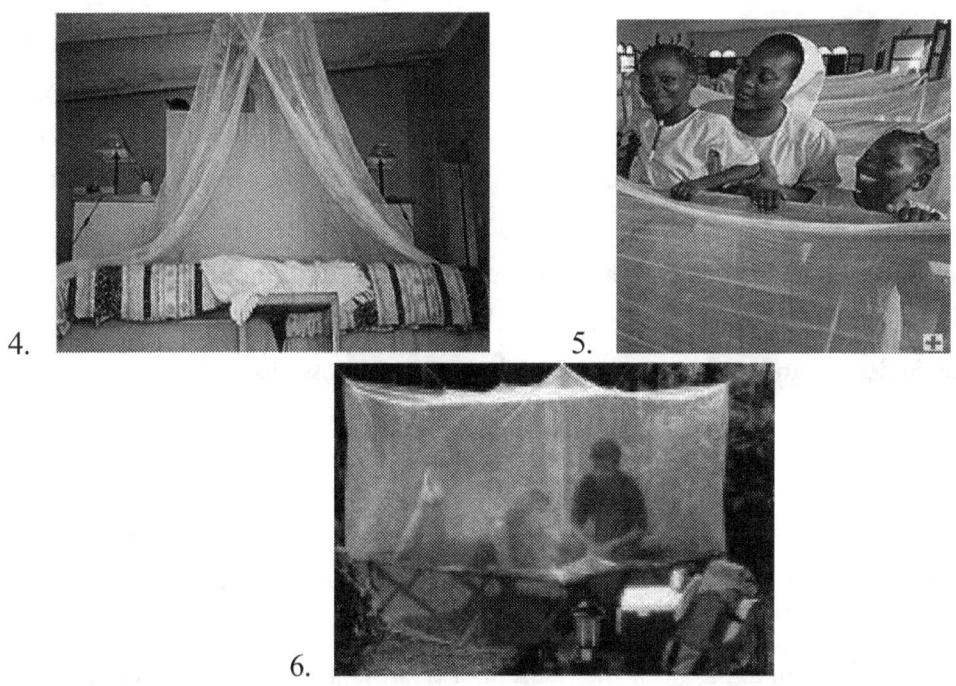

4. 5.

6.

Figures 4: The use of ITNs is shown by coverage over a bed respectively.
Figure 5: Orphans in a joyous mood after receiving ITNs to help keep them from malaria infection.
Figure 6: A family enjoying the sunny African weather and staying protected as well by confining themselves using mosquito nets.

With regards to controlling the widespread of the disease, we can implement three major techniques which have been recommended by the World Health Organization. First is the application of vector control selectively. This may involve reducing the population of vector mosquitoes by termination or destroying their breeding grounds which results in the elimination of their life cycle; i.e., destroying their complete morphological cycle which also consist of the larval, pupal and stages. By doing this, there is definitely going to be suppression in the contact between the vectors (mosquitoes) and their victims (humans). This technique is the most important when viewed from an environmental perspective. The reason is because environmentally we can consider the flushing of irrigation channels, irrigating stagnant waters, keeping water containers unexposed and the introduction of tilapia into ponds since they can also feed on the eggs and larvae of mosquitoes. These factors are very efficient when it comes to destroying the vector's breeding sites (10). Other related methods of chemical control includes the casting of petroleum oils onto stagnant waters as this causes oxygen starvation for the larvae and pupae thus resulting in their deaths and incompletion of their morphological development into adult mosquitoes which is responsible for the widespread of malaria. Also is the use of copper acetoarsenite. This larvicide, particle like in nature, is ingested by the larvae and as a result of it being poisonous, lead to their eliminations as well. (11).

Adjacently, is the use of biological control. By way of example we can look at two types of bacteria which have been used over the past decade, namely *Bacillus sphaericus* and *Bacillus thuringiensis*. (12, 13). What happens is during sporulation, the toxic bacteria produce as crystals protoxins affects the larvae after ingestion. These protoxins being solubilized in the alkaline pH of the larval midgut are activated proteolytically. Afterwards, they bind to the epithelial cells' specific receptors which lead to cell lysis and the larva stops feeding and perishes.

Secondly, would be the diagnosis as well as the effective and prompt treatment for malaria so as to minimize the rate of morbidity and mortality. The final technique can imply the detection of malaria epidemics and rapid control methods as early as possible.

Malaria Treatment and Resistance

As malaria has plagued sub-Saharan Africa for almost a century, different drugs have been used to combat this disease caused by the infections of P. falciparum, P. ovale, P. vivax, and P. malariae.

Primarily, chloroquine has been the household treatment over the past decades. Even after 1979, when the first case of it being resisted by malaria (P. falciparum) was reported, the drug was, and as a matter of fact, is still being used by victims simply because of its cheap cost (10 cents per curative regimen) and affordability. As compared to other drugs which are more costly, than chloroquine, has become highly ineffective against the falciparum parasite (3, 5).

Immunologically, the reason behind chloroquine being resistant by the said parasite is because chloroquine acts by binding to heme molecules released from the hemoglobin. As these molecules are eventually digested by the malaria parasite, they grow within the erythrocytes (RBCs) of their host. This binding interferes with the process by which heme is normally incorporated into inert crystals and detoxified, thereby effectively poisoning the parasite. Because heme is not a parasite-encoded molecule that can mutate under drug pressure, malaria-fighting producers have had to solve the difficult problem of chloroquine's toxicity by evolving a mechanism either to prevent drug-heme interactions or to control the damage from these complexes. P. falciparum achieves this by reducing chloroquine accumulation in their acid digestive vacuoles where the drug does its damage (5).

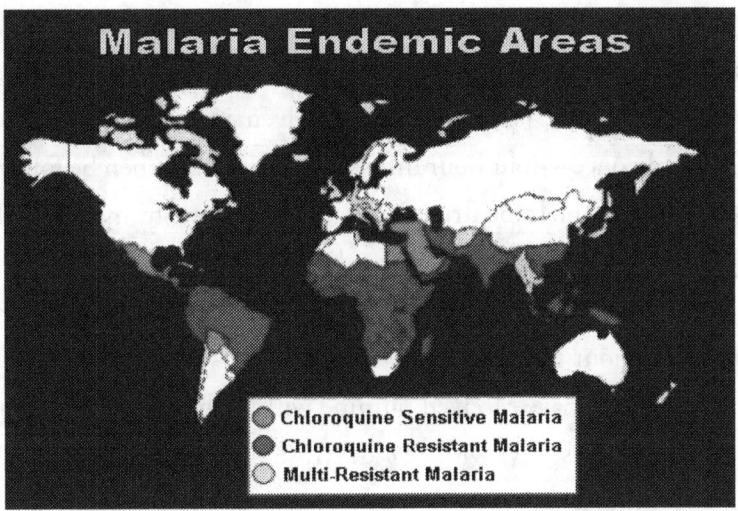

Since the four groups of malaria causing parasites exist in different areas, we can separate them and view a variety of efficient treatments, which will meet no resistance by the parasites involved (1).

(I) P. falciparum:

With this group, malaria can exist as uncomplicated or complicated.

A.   Uncomplicated Malaria (oral therapy) can be treated using one of three regimens.

1.   Quinine sulfate (salt), 10 mg/ 8 kg hourly for 7 days plus 100 mg of doxycycline daily for a week as well. Since doxycycline is a little costly, 4 mg/ kg of tetracycline can be taken for the same period, or a single fansidar dose containing 25 mg/ kg of sufladoxine plus 1.25 mg/ kg pyramethamine can be taken. Side effects usually involve the development of "cinochonism" (tinnitus, high tone hearing loss, nausea and dysphoria) after two or three days. NOTE: Patient should continue to avoid recrudescence.

2.   4 Malarone (atovaquone 250 mg plus proguaril 100 mg) tablets daily for 3 consecutive days. As compared to other treatments, this therapy is very expensive.

3.   15 mg/ kg of Mefloquine (Larium) in a divided dose followed by 10 mg/ kg the next day. To avoid nauseation, antipyretic and antemetic agents (Dolasetron [Anzemet], Aprepitant [Emend], etc) can be given prior to the drug administration.

Also Lapdap (2 mg/ kg chlorproguanil plus 2.5 mg/ kg of dapsone) can be taken for three consecutive days. It is very affordable and perfect for sub-Saharan Africa.

B.   Severe Malaria (involving coma, jaundice, renal failure, high parasitic counts, sever anemia, etc) is treated intensively since patients have to be monitored closely-clinically and biochemically. Intravenous quinine is the best treatment but rapid injection can be detrimental (1).

After 48 hours before care is given, patents can receive one or two regimens:

1. Quinine dihydrochloride (salt), 20 mg/ kg given intravenously, with 5% dextrose or normal saline only once—four hour infusion. This should then be followed by four hours later by 10 mg/ kg of the same drug using the same time frame but eight hourly.

2. During syringe use, the above drug can be administered by using 7 mg/ kg over 30 minutes followed by an increment of 3 mg/ kg over four hours, then starting four hours later again as four hour infusion, eight hourly.

Note: As for patients who have received quinine within the previous 24 hours, 10 mg/kg of dihydrochloride salt base I.V. with 5% dextrose or normal saline as 4 hour infusion, 8 hourly can be administered.

(II) P.Vivax:

Most strains of *P. vivax* are still sensitive to chloroquine. This drug will clear the erythrocyte stages of the parasite but it has no effect on the exo-erythrocytic liver stage and a course of primaquine (an 8-amino-quinoline) is required for radical cure (1). The Chesson strain of *P. vivax* shows some resistance to primaquine and an increased dose is required but may still not result in radical cure. If primaquine is not given, the patient may suffer a relapse which will occur weeks, months or sometimes years after the original attack. The primaquine is preferably started after the chloroquine, at a dose of 3.5mg/kg given as a divided daily dose over 14 days. An effective way of ensuring the treatment is to first take chloroquine tablet for 3 days in a 6-2-2 manner for days 1-3 respectively (1). For the next 14 days, 2 primaquine tablets can be taken per day. For those with allergic to chloroquine, malarone can be substituted, but taken along with chloroquine. In the case of a relapse, repeat both chloroquine and primaquine treatment. Several relapses may occur before the parasite is finally eliminated. Unfortunately there is no other effective treatment. Tafenoquine is a newly developed long-acting, potent primaquine-like drug which may be available soon for radical cure. It can be given over 3 days instead of 14 but may not offer any advantage in efficacy over primaquine (1).

Patients should have their G6PD (Glucose-6-phosphate dehydrogenase) status checked before primaquine (or tafenoquine) is prescribed. Those with G6PD deficiency may experience haemolysis if given a daily dose of primaquine and it is recommended that these patients be given 30-45mg once a week for 8 weeks (1).

(III) P. malariae, P. ovale:

Treatment for the eradication of these two strains of malaria is the same as that for *P.*

vivax except it is not necessary to give primaquine to those patients with *P. malariae*(1).

Artemisinin-based Combination Therapy

Artemisinin has been used for many years by the Chinese as a traditional treatment for fever and malaria. It is a sesquiterpene lactone derived from the wormwood plant *Artemisia annua*. Semi-synthetic derivatives including artemether and artesunate are now widely available in the tropics. These compounds are being increasingly used in a number of countries and are both cheap and effective. They are starting to be licensed in Western countries. (4).

They are particularly valuable in the treatment of multidrug-resistant falciparum malaria. Unless used with a second antimalarial as described below there is likely to be a high recrudescent rate. Side-effects have been reported but these are comparartively rare and seldom severe. Artemisinin derivatives are recommended for treatment but not for prophylaxis. If an artemisinin drug is used to treat vivax malaria, it should be accompanied by a course of primaquine. (4).

Artemisinin (500mg tablets) give 10-20 mg/kg on day 1 (500-1,000 mg) orally then 500mg for 4 days. Then give mefloquine 15mg base/kg or split dose 25mg base/kg.

Artemisinin (200mg suppositories): for **severe malaria** 600-1200mg stat, 400-600mg after 4 hours then 400-800mg twice daily for 3 days. Give mefloquine as above.

Artesunate (50 & 60 mg vials for intravenous use): for **severe malaria** 120mg I.V. stat. 60 mg at 4, 24 and 48 hours, 50-60 mg on days 3-5. Give mefloquine as above.

Dihydroartemisinin (20 mg tablets): First dose 120mg then 60mg daily for 4-6 days then give mefloquine as above.

Artemeter (vials for intramuscular use): For **severe malaria** 3.2 mg/kg intramuscularly stat then 1.6mg twice daily for 3-7 days, give mefloquine as above.

Caution: A recent report on the sale of artesunate in South East Asia found that 38% of the artesunate purchased was fake (8). Fake artesunate can be identified by a dye test (1). Other fake drugs including mefloquine can also be found in South East Asian markets.

Artemisinin-based combination therapies, some comprising a fixed co-formulation are being developed:

Artesunate plus mefloquine is used in several Asian countries, some using blister packs to simplify dosing and increase compliance. The mefloquine component, which is usually given on the third day of artesunate treatment, is relatively expensive and side-effects are common.

Due to drug related complications, one must always remember to inform the pharmacist or pharmacy technician during the time of purchase about known allergic reactions, other drugs being taken and if you are experiencing any other medical condition(s). Also pregnant mothers, women planning a pregnancy and breast feeders should have the pharmacist informed as well. This information in particular is susceptible to prevent any medical mishap to the foetus or infant. As for travellers, it is important to purchase and take malaria preventive drugs at least two to three weeks before the departure date. One must be sure to purchase the right preventive drug since the infection is carried out by different species of the Plasmodium parasite as this will ensure its effectiveness. Finally, once abroad, it must be made an unforgettable rule to always take the drugs until the individual returns home. It is also advisable to continue treatment for about a week just to insure that your system is fully cleared.

A Case Study in Liberia

Geography

Liberia is situated on the west coast of Africa between 4 degrees 32 minutes and 8 degrees 50 minutes longitude west of Greenwich. Liberia is about 250 miles above the Equator and is bounded in the north by the Republic of Guinea, in the east by La Cote d'Ivoire, in the west by Sierra Leone and in the south by the Atlantic Ocean. The land covers a total area of 43,000 square miles and half of this land area is covered by evergreen high forests. The terrain rises in four distinct steps, each both running roughly parallel to the coastline. The coastal plain itself is between 10 and 25 miles wide. It is an area of extreme variety, where patches of forest mingle with scrub and thorn, where mangrove swamps rub shoulders with cultivated land.

Climate:

Liberia has a tropical climate with average temperatures around 70 degrees Fahrenheit (21 degrees Celsius). There are relatively small variations between day and night and between seasons, with temperatures rarely exceeding 100 degrees Fahrenheit (37 degrees Celsius).

There are two seasons – the rainy season which is from May to October and the dry from November to April. It seldom rains during the dry season, though there are dry period during the rainy season, including a dry spell in July or August lasting about two weeks. The annual rainfall averages 170 inches (4.320 mm.) inland. The average humidity on the coastal belt is 78% during the wet season, but it is liable to drop to 30% from December to March when the Harmattan winds blow from the Sahara.

Background of Liberia :

Liberia has a unique history among African states. It was originally established as a colony by the philanthropic National Colonization Society of America as a homeland for freed American slaves. Land was obtained by treaty with indigenous tribes in 1822, and was subsequently settled. In 1847, Liberia proclaimed itself an independent republic, but failed to live up to the name. Later generations of American Liberians established a slave-state of their own, enslaving the indigenous Liberians. (6)

Slavery continued until the 1936, when it was abolished in the wake of scandal and League of Nations investigation. The scandal did not manage to unseat the True Whig party from its then more than half-century of domination of Liberian politics.

Liberia continued as a republic under the Whigs until April of 1980, when Mast.Sgt. Samuel K. Doe and his People's Redemption Council staged a bloody coup. The PRC established itself as an interim regime until 1985, when elections were held. Doe and the PRC were returned to power among allegations of election fraud. (6)

On December 24, 1989, a small band of rebels led by Doe's former procurement chief, Charles Taylor, invaded Liberia from the Ivory Coast. Taylor and his National Patriotic Front rebels rapidly gained the support of Liberians because of the repressive nature of Samuel Doe and his government. Barely 6 months after the rebels first attacked, they reached the outskirts of Monrovia. (6)

The 1989-1996 Liberian civil war, which was one of Africa's bloodiest, claimed the lives of more than 200,000 Liberians and further displaced a million others into refugee camps in neighboring countries. The Economic Community of West African States (ECOWAS) intervened and succeeded in preventing Charles Taylor from capturing Monrovia. Prince Johnson--who had been a member of Taylor's National Patriotic Front of Liberia (NPFL) but broke away because of policy differences--formed the Independent National Patriotic Front of Liberia (INPFL). Johnson's forces captured and killed Doe on September 9, 1990. (6)

After considerable progress in negotiations conducted by the United States, United Nations, Organization of African Unity (now the African Union), and ECOWAS, disarmament and demobilization of warring factions were hastily carried out. Special elections were held on July 19, 1997, with Charles Taylor and his National Patriotic Party emerging victorious. Taylor won the election by a large majority, primarily because Liberians feared a return to war had Taylor lost. (6)

For the next 6 years, the Taylor government failed to better the lives of Liberians. Unemployment and illiteracy stood above 75%, and little investment was made in the country's infrastructure. Liberia is still trying to recover from the ravages of war; six years after the war, pipe-borne water and electricity were still unavailable, and schools, hospitals, roads, and infrastructure remained derelict. Rather than work to improve the lives of Liberians, Taylor supported the bloody Revolutionary United Front in Sierra Leone, fomenting unrest and brutal excesses in the region, and leading to the resumption of armed rebellion from among Taylor's former adversaries. (6)

On June 4, 2003 in Accra, Ghana, ECOWAS facilitated the inauguration of peace talks among the Government of Liberia, civil society, and the rebel groups called "Liberians United for Reconciliation and Democracy" (LURD) and "Movement for Democracy in Liberia" (MODEL). LURD and MODEL largely represent elements of the former ULIMO-K and ULIMO-J factions that fought Taylor during Liberia's previous civil war (1989-1996). Also on June 4, 2003, the Chief Prosecutor of the Special Court for Sierra Leone issued a press statement announcing the opening of a sealed March 7 indictment of Liberian President Charles Taylor for "bearing the greatest responsibility" for atrocities in Sierra Leone since November 1996. By July 17, 2003 the Government of Liberia, LURD, and MODEL signed a cease-fire that envisioned a comprehensive peace agreement within 30 days. The three combatants subsequently broke that cease-fire repeatedly, which resulted in bitter fighting that eventually reached downtown Monrovia. (6)

On August 11, 2003 under intense U.S. and international pressure, President Taylor resigned office and departed into exile in Nigeria. This move paved the way for the deployment by ECOWAS of what became a 3,600-strong peacekeeping mission in Liberia (ECOMIL). Since then, the United States has provided limited direct military support and $26 million in logistical assistance to ECOMIL and another $40 million in humanitarian assistance to Liberia. On August 18, leaders from the Liberian Government, the rebels, political parties, and civil society signed a comprehensive peace agreement that laid the framework for constructing a 2-year National Transitional Government of Liberia, effective October 14. On August 21, they selected businessman Gyude Bryant as Chair and Wesley Johnson as Vice Chair of the National Transitional Government of Liberia (NTGL). Under the terms of the agreement the LURD, MODEL, and Government of Liberia each selected 12 members of the 76-member Legislative Assembly (LA). The NTGL was inducted on October 14, 2003 and will serve until January 2006, when the winners of the scheduled October 11, 2005 presidential and congressional elections take office. The election was won by Mrs. Ellen Johnson- Sirleaf after going through a second round stand-off with soccer legend George Oppong Weah.(6)

As Liberia is now trying to once again show her perpendicularity, she is taking a prominent step in the health sector. Due to more than a decade of civil fighting, malaria became highly endemic in the country and one of the major public health problems. At health centers, the disease is responsible for 40-45 % at OPDs (Out Patient Departments) as well as inpatient death by 17.8%. (6).

Liberia's War on Malaria

Currently, as Liberia is still under sanctions due to the just ended civil crisis, accessibility due to insecurity has caused the unavailability of proper medication and control tools (insecticides, etc) to the population.

From a socio-economical standpoint, the cost of treatment to families versus the cost of unemployment as well as the maintenance of pre-war salary rate is very high. Because of this, families have over the past decade relied on chloroquine, the cheapest treatment available despite its growing resistance by malaria which is currently above 74 % was first noted in 1988- a year before the outbreak of the country's civil crisis. In addition to this, vector control at the household level has also been impossible due to ignorance and poverty as well. (4).

As a declaration was recently signed in Abuja, Nigeria by African leaders with a commitment to reduce the mortality rate by malaria on the continent to 50% by 2010, Liberia has decided to work parallel with the declaration. Being dedicated to the program, she launched a 4 year (2004-08) "National Strategic Plan" (NSP) which is in uniformity with the World Health Organization's (WHO) policy on rolling back malaria.

The "Roll Back Malaria" campaign is expected to have 3 affirmative outcomes:

1. The burden of disease being reduce by 50%
2. Human development
3. Poverty reduction, all of which is being supported by financial institutions like the Global Funds which addresses the malaria epidemic.

Liberia's overall objective until 2010 is to reduce morbidity and mortality by 50%. This idea of suppression rather than eradication serves as a preliminary step due to the country's recovery from a civil crisis. (6).

As illiteracy has served as a major catalyst for the widespread of malaria in this West African republic, Information, Education and Communication (IEC) of the Ministry of Health (MOE) has decided to step up awareness campaigns on the issue of malaria. In conjunction with this institution, the Ministry of Education and other media sectors are joining forces to insure that the message is absorbed and assimilated throughout the general public. (7).

With this plan in mind, the year 2008 has been targeted to have the following achieved:

- 80% of the population nationwide is to be fully aware of malaria.
- 80% of community health worker, care-givers, and drug store sellers are to be enlightened on the dissemination of malaria related information.

- 50% of households are to receive IEC and Behavior Change Communications (BCC) and therefore change their poor health practices.
- 90% of schools should receive IEC components on the disease as well as controlling and preventing it.

As a means of treating this disease currently, there is the implementation of the Artemisinin-base Combination Therapy (ACT) and alternatively a combinative dosage of chloroquine and sulphadoxine pyrimethamine (SP). (8). On the other hand, preventive methods include:

1. People being educated and encouraged on the usage of insecticide treated nets (ITNs) as the population's knowledge was found to be very low.

2. The use of Intermittent Preventive Treatment (IPT) such as SP by pregnant women which lowers the rate of fetal infection and mortality (still birth) is being encouraged. (8).

3. Increasing the combination of personal and community protective measures among those at risk-Keeping the environment clean by terminating breeding grounds use by the vector of this disease. (7). These consist of clogged drainages, polluted rivers and creeks, water-logged sites as well as thick bushes all within the communities and their surroundings.

Some polluted and mosquito infested sites around the capital city of Monrovia:

An exposed dump site around the U.N. Drive area (A major street in central Monrovia. As one will agree, this exposes the community to a high risk of malaria infection as this site serves as a healthy breeding ground for mosquitoes).

A community also within the Monrovia area which can be suspected of malaria infection alone due to the clogged drainage that runs through it.
The pictures below show an example of how communities within the country are at high risk of malaria infection

Another community within the Monrovia area which can be suspected of malaria infection alone due to the clogged drainage that runs through it.

Another perfect breeding ground for mosquitoes.

A high level of pollution engulfs the Waterside shopping center as people go about their normal business on the Gabriel Tucker's bridge hoping that the government will exterminate this eye sol.

Conclusion

After reading this source of enlightenment, we can all agree that this form of terrorism in disguise has unfolded its greatest effect on the African continent and inconspicuously on other parts of the tropics. It is highly imperative that countries work together for the universal suppression and eradication of malaria. This measure is definitely going to result in the affirmative by lowering infection and reducing human death (2.7 million die of the disease per annum). Ensuring the combat against malaria can be executed by having training workshops so that individuals can be trained and qualified to disseminate the message on the disease and its harmful effects to the general public. This initial step can be launched by using the media, schools, etc. Also, local governments can help enforce laws on the behavior change of poor health practitioners. To attach firmness to this act, violators can be fined for their wrong doings or prosecuted to the fullest extent depending on the gravity of the offense. This is especially for those who contribute to the development of breeding grounds for the vector of this disease.

Finally, once the proper preventive measures are taken, we can expect malaria to decline since the mosquito population as a result will then be under control.

"There is light behind darkness"

Acknowledgements

I would like to give thanks and Praises to God Almighty for giving me the wisdom, strength and courage to embark and complete this prestigious task.

I would like to thank my parents for their priceless efforts and support that enabled me to acquire facts and data needed from Liberia. Also I would like to give special thanks to Dr. Florence Okafor (Prof. of Biology, Alabama A & M University) for her endless moral and academic guidance that were paramount in steering my ability to accomplish this task. May the Good Lord bless you all for your contributions.

References

1. J Pharm Biomed Anal 2000; 24: 65-70

2. Karch S, Asidi N, Manzambi Z et al
    J Am Mosq Control Assoc. 1995 June; 11(2.1): 191-194

3. Kamya M.R., Bakyaita N.N. et al
    Increasing antimalarial drug resistance in Uganda and revision of the nat'l drug policy.
    Tropical Medicine and int'l Health
    Vol. 7 Issue 12, pg 1031, December 2002.

4. Trape, Jean-Francoise. "The Public Health Impact of Chloroquine Resistance in Africa."
    The American Journal of Tropical Health and Hygiene. 2001; 64 (1, 2): 12-17.

5. Wellems, T.E. 2002. Plasmodium Chloroquine Resistance and the Search for a
    Replacement Antimalarial Drug. Science, vol. 298; pp. 124-125.

6. MCD: Routine Malaria Surveillance Data: 1993- 1999

7. Freeman, T.L. & Bolay, F.T (1995). In vivo response of Plasmodium falciparum to standard
    chloroquine regimen in Buchanan, Grand Bassa County, Liberia (Unpublished)

8. Checchi, F et al (2002). High Plasmodium falciparum resistance to chloroquine and
    sulphadoxine- pyrimethamine in Harper, Liberia: results in vivo and analysis of point
    mutation.

9. Hoffman, Stephen L; Subramanian, G. Mani; et al. "Plasmodium, human and Anopheles
    genomics and malaria." Nature. 2002, February; 415, 702-709

10. World Health Organization.1995. Vector control for malaria and other mosquito- borne
    diseases. WHO Tech. Rep. Ser. 857:1- 91

11. Bruce- Chwatt, L. 1993. In H.M Gillies and D.A. Warrall (ed.), Essential malariology,
    3rd ed. Edward Arnold, London, United Kingdom.

12. Porter, A. G. 1996. Mosquitocidal toxins, genes and bacteria: the hit squad. Parasitol.
    Today 12: 175-179.

13. Priest, F. G. 1992. Biological control of mosquitoes and other biting flies by Bacillus
    sphaericus and Bacillus thuringiensis. J. App. Bacteriol. 72: 357-369.

14. Related Websites:
    A. www.netdoctor.co.uk

    B. www.traveldoc.info/disease

    C. www.malariasite.com

    D. www.cdc.gov/malaria/disease.htm

    E. www.cdc.gov/mamaria/diagnosis_treatment/treatment.htm

    F. www.cdc.gov/malaria/diagnosis_treatment/diagnosis.htm

    G. www.mmv.org

    H. www.sciencemag.org

    I. www.bact.wisc.edu/microtext book/

About the Author

Born unto the union of Mr. Rockefeller F. Cooper I and Mrs. Delrene H-Cooper in the West African Republic of Liberia, was a son they named Rockefeller F. Cooper II. Growing up as a child, he started and earned his primary education at the School of Prime System and later started his elementary education at the J.J. Roberts United Methodist School. After completing, he started his junior high education at the same school which came to an abrupt stop due to a deadly civil war in the hinterlands of the country which was encroaching upon the capital city of Monrovia. As he and his family decided to migrate into the rural parts of the country in order to seek safety, they finally sorted refuge in Suacaco, Bong County on the campus of the Cuttington University College which at the time was abandoned and serving as a refuge site cum internally displaced people. Still eager to learn and undeterred by the current situation, he enrolled at Cuttington Community School at the birth of a ceasefire agreement between rebel forces and peace keeping forces (ECOMOG). After a year, Rockefeller and his family moved to Buchanan, Grand Bassa County where he completed junior high and started his senior high schooling at the St. Peter's Claver High School.

Unfortunately, as fighting reignited intensively in 1992, the Cooper family saw it necessary to flee the country and in 1993 settled in Ghana. Rockefeller then enrolled at the Takoradi Secondary School whiles viewing life from a different perspective. Since the country's education system was based on that of the British due to colonial ties, he decided to concentrate his studies in the natural sciences. This decision came to mind after seeing all the unnecessary destruction of human life and thus having the urge to serve humanity by one day becoming a medical physician. Upon completing his course of study in 1996, he returned to his homeland the following year after the country had its first post-war elections. In October of 1998, he started his college education at the Cuttington University College as biology major. In 2001, he transferred to the Alabama A&M University where he completed his bachelor's degree with honors (Magna Cum Laude) by

May of 2003. Currently, Rockefeller is a graduate student at the same university where he hopes to earn a master's degree in the field of biology as well by the Summer of 2006.